The Standard

FLIGHT CREW LOG

Trip and Expense Record

ASA-SP-FC

Flight Crew Log
SP-FC

ASA-SP-FC
ISBN 978-1-56027-400-1

Published by
Aviation Supplies & Academics, Inc.
7005 132nd Place SE
Newcastle, Washington 98059-3153
Website: www.asa2fly.com
Email: asa@asa2fly.com

Printed in the United States of America

21/2019

TRANSPORTATION USD $6.95

ISBN 978-1-56027-400-1

FLIGHT CREW LOG

Name ____________________

Address ____________________

____

Telephone ____________________

Book No. ____________________

From Date __________ Hours __________

To Date __________ Hours __________

Pilot Certificate No. ____________________

Company ____________________

Base ____________________

Position ____________________

Seniority Number ____________________

	Date Issued	Date Due
Currency		
Checkride		
Medical		
________		
________		

Month ____________ Year ____________ Line No. ____________

Sunday	Monday	Tuesday	Wednesday	Thursday	Friday	Saturday

Month ______________ Year ______________ Line No. ______________________________

Sunday	Monday	Tuesday	Wednesday	Thursday	Friday	Saturday

Month ______________ Year ______________ Line No. ______________

Sunday	Monday	Tuesday	Wednesday	Thursday	Friday	Saturday

Month ______________ Year ______________ Line No. ______________________________

Sunday	Monday	Tuesday	Wednesday	Thursday	Friday	Saturday

Month ______________ Year ______________ Line No. ______________

Sunday	Monday	Tuesday	Wednesday	Thursday	Friday	Saturday

Month ______________ Year ______________ Line No. ______________________________

Sunday	Monday	Tuesday	Wednesday	Thursday	Friday	Saturday

Month ____________ Year ____________ Line No. ______________________________

Sunday	Monday	Tuesday	Wednesday	Thursday	Friday	Saturday

Month ______ Year ______ Line No. ______

Sunday	Monday	Tuesday	Wednesday	Thursday	Friday	Saturday

Month ______ Year ______ Line No. ______

Sunday	Monday	Tuesday	Wednesday	Thursday	Friday	Saturday

Month ______________ Year ______________ Line No. ______________

Sunday	Monday	Tuesday	Wednesday	Thursday	Friday	Saturday

Month ______________ Year ______________ Line No. ______________

Sunday	Monday	Tuesday	Wednesday	Thursday	Friday	Saturday

Month ______________ Year ______________ Line No. ______________

Sunday	Monday	Tuesday	Wednesday	Thursday	Friday	Saturday

Date	Flight No.	A/C Type	A/C ID	From	To	Time			Flight Conditions				
						Dep	Arr	TOTAL	Inst	App	Night		

Duty Start ______ Int'l Pay ______ Page Total ___ ___ ___ ___ ___ ___

Duty Stop ______ Night Pay ______ Amt Fwd ___ ___ ___ ___ ___ ___

Total Duty ______ Pay ______ Monthly Total ___ ___ ___ ___ ___ ___

Remarks: ______

Expenses Hours ______

Breakfast	Lunch	Dinner	Snack	Hotel	Phone	Taxi	Cleaning	Tips	Misc.	TOTAL

Trip Seq. No. ______ Page Total ______

Crew Members ______ Amount Forward ______

Monthly Total ______

Date	Flight No.	A/C Type	A/C ID	From	To	Time			Flight Conditions				
						Dep	Arr	TOTAL	Inst	App	Night		

Duty Start______ Int'l Pay______ Page Total______

Duty Stop______ Night Pay______ Amt Fwd______

Total Duty______ Pay______ Monthly Total______

Remarks:______

Expenses Hours______

Breakfast	Lunch	Dinner	Snack	Hotel	Phone	Taxi	Cleaning	Tips	Misc.	TOTAL

Trip Seq. No.______ Page Total______

Crew Members______ Amount Forward______

Monthly Total______

Date	Flight No.	A/C Type	A/C ID	From	To	Time			Flight Conditions				
						Dep	Arr	TOTAL	Inst	App	Night		

Duty Start________ Int'l Pay________ Page Total______ ______ ______ ______ ______ ______

Duty Stop________ Night Pay________ Amt Fwd______ ______ ______ ______ ______ ______

Total Duty________ Pay________ Monthly Total______ ______ ______ ______ ______ ______

Remarks:________________________________

____

Expenses Hours________

Breakfast	Lunch	Dinner	Snack	Hotel	Phone	Taxi	Cleaning	Tips	Misc.	TOTAL

Trip Seq. No.________________ Page Total________

Crew Members________________________ Amount Forward________

________________________ Monthly Total________

Date	Flight No.	A/C Type	A/C ID	From	To	Time			Flight Conditions				
						Dep	Arr	TOTAL	Inst	App	Night		

Duty Start______ Int'l Pay______ Page Total______ ______ ______ ______ ______ ______

Duty Stop______ Night Pay______ Amt Fwd______ ______ ______ ______ ______ ______

Total Duty______ Pay______ Monthly Total______ ______ ______ ______ ______ ______

Remarks:______

Expenses Hours______

Breakfast	Lunch	Dinner	Snack	Hotel	Phone	Taxi	Cleaning	Tips	Misc.	TOTAL

Trip Seq. No.______ Page Total______

Crew Members______ Amount Forward______

Monthly Total______

Date	Flight No.	A/C Type	A/C ID	From	To	Time			Flight Conditions				
						Dep	Arr	TOTAL	Inst	App	Night		

Duty Start ______ Int'l Pay ______ Page Total ______ ______ ______ ______ ______ ______

Duty Stop ______ Night Pay ______ Amt Fwd ______ ______ ______ ______ ______ ______

Total Duty ______ Pay ______ Monthly Total ______ ______ ______ ______ ______ ______

Remarks: ______

______

Expenses Hours ______

Breakfast	Lunch	Dinner	Snack	Hotel	Phone	Taxi	Cleaning	Tips	Misc.	TOTAL

Trip Seq. No. ______ Page Total ______

Crew Members ______ Amount Forward ______

______ Monthly Total ______

Date	Flight No.	A/C Type	A/C ID	From	To	Time			Flight Conditions				
						Dep	Arr	TOTAL	Inst	App	Night		

Duty Start________ Int'l Pay________ Page Total____ ____ ____ ____ ____ ____

Duty Stop________ Night Pay________ Amt Fwd____ ____ ____ ____ ____ ____

Total Duty________ Pay________ Monthly Total____ ____ ____ ____ ____ ____

Remarks:________________________________

Expenses Hours________

Breakfast	Lunch	Dinner	Snack	Hotel	Phone	Taxi	Cleaning	Tips	Misc.	TOTAL

Trip Seq. No.____________ Page Total________

Crew Members____________________ Amount Forward________

Monthly Total________

Date	Flight No.	A/C Type	A/C ID	From	To	Time			Flight Conditions				
						Dep	Arr	TOTAL	Inst	App	Night		

Duty Start________ Int'l Pay________ Page Total____ ____ ____ ____ ____ ____

Duty Stop________ Night Pay________ Amt Fwd____ ____ ____ ____ ____ ____

Total Duty________ Pay________ Monthly Total____ ____ ____ ____ ____ ____

Remarks:________________________________

Expenses Hours________

Breakfast	Lunch	Dinner	Snack	Hotel	Phone	Taxi	Cleaning	Tips	Misc.	TOTAL

Trip Seq. No.________ Page Total________

Crew Members________ Amount Forward________

Monthly Total________

Date	Flight No.	A/C Type	A/C ID	From	To	Time			Flight Conditions				
						Dep	Arr	TOTAL	Inst	App	Night		

Duty Start______ Int'l Pay______ Page Total____ ____ ____ ____ ____ ____

Duty Stop______ Night Pay______ Amt Fwd____ ____ ____ ____ ____ ____

Total Duty______ Pay______ Monthly Total____ ____ ____ ____ ____ ____

Remarks:______

Expenses Hours______

Breakfast	Lunch	Dinner	Snack	Hotel	Phone	Taxi	Cleaning	Tips	Misc.	TOTAL

Trip Seq. No.______ Page Total______

Crew Members______ Amount Forward______

Monthly Total______

Date	Flight No.	A/C Type	A/C ID	From	To	Time			Flight Conditions				
						Dep	Arr	TOTAL	Inst	App	Night		

Duty Start______ Int'l Pay______ Page Total______

Duty Stop______ Night Pay______ Amt Fwd______

Total Duty______ Pay______ Monthly Total______

Remarks:______

Expenses Hours______

Breakfast	Lunch	Dinner	Snack	Hotel	Phone	Taxi	Cleaning	Tips	Misc.	TOTAL

Trip Seq. No.______ Page Total______

Crew Members______ Amount Forward______

Monthly Total______

Date	Flight No.	A/C Type	A/C ID	From	To	Time			Flight Conditions				
						Dep	Arr	TOTAL	Inst	App	Night		

Duty Start________ Int'l Pay________ Page Total____ ____ ____ ____ ____ ____

Duty Stop________ Night Pay________ Amt Fwd____ ____ ____ ____ ____ ____

Total Duty________ Pay________ Monthly Total____ ____ ____ ____ ____ ____

Remarks:________________________________

____

Expenses Hours________

Breakfast	Lunch	Dinner	Snack	Hotel	Phone	Taxi	Cleaning	Tips	Misc.	TOTAL

Trip Seq. No.____________ Page Total________

Crew Members____________________ Amount Forward________

____________________________ Monthly Total________

Date	Flight No.	A/C Type	A/C ID	From	To	Time			Flight Conditions				
						Dep	Arr	TOTAL	Inst	App	Night		

Duty Start________ Int'l Pay________ Page Total____ ____ ____ ____ ____ ____

Duty Stop________ Night Pay________ Amt Fwd____ ____ ____ ____ ____ ____

Total Duty________ Pay________ Monthly Total____ ____ ____ ____ ____ ____

Remarks:________________________________

Expenses Hours________

Breakfast	Lunch	Dinner	Snack	Hotel	Phone	Taxi	Cleaning	Tips	Misc.	TOTAL

Trip Seq. No.____________ Page Total________

Crew Members____________________ Amount Forward________

Monthly Total________

Date	Flight No.	A/C Type	A/C ID	From	To	Time			Flight Conditions				
						Dep	Air	TOTAL	Inst	App	Night		

Duty Start______ Int'l Pay______ Page Total______

Duty Stop______ Night Pay______ Amt Fwd______

Total Duty______ Pay______ Monthly Total______

Remarks:______

Expenses Hours______

Breakfast	Lunch	Dinner	Snack	Hotel	Phone	Tax	Cleaning	Tips	Misc.	TOTAL

Trip Seq. No.______ Page Total______

Crew Members______ Amount Forward______

Monthly Total______

Date	Flight No.	A/C Type	A/C ID	From	To	Time			Flight Conditions				
						Dep	Arr	TOTAL	Inst	App	Night		

Duty Start________ Int'l Pay________ Page Total____ ____ ____ ____ ____ ____

Duty Stop________ Night Pay________ Amt Fwd____ ____ ____ ____ ____ ____

Total Duty________ Pay________ Monthly Total____ ____ ____ ____ ____ ____

Remarks:________________________

Expenses Hours________

Breakfast	Lunch	Dinner	Snack	Hotel	Phone	Taxi	Cleaning	Tips	Misc.	TOTAL

Trip Seq. No.____________ Page Total________

Crew Members____________________ Amount Forward________

Monthly Total________

Date	Flight No.	A/C Type	A/C ID	From	To	Time			Flight Conditions				
						Dep	Arr	TOTAL	Inst	App	Night		

Duty Start______ Int'l Pay______ Page Total______

Duty Stop______ Night Pay______ Amt Fwd______

Total Duty______ Pay______ Monthly Total______

Remarks:______

Expenses Hours______

Breakfast	Lunch	Dinner	Snack	Hotel	Phone	Taxi	Cleaning	Tips	Misc.	TOTAL

Trip Seq. No.______ Page Total______

Crew Members______ Amount Forward______

Monthly Total______

Date	Flight No.	A/C Type	A/C ID	From	To	Time			Flight Conditions				
						Dep	Arr	TOTAL	Inst	App	Night		

Duty Start________ Int'l Pay________ Page Total________

Duty Stop________ Night Pay________ Amt Fwd________

Total Duty________ Pay________ Monthly Total________

Remarks:________

Expenses Hours________

Breakfast	Lunch	Dinner	Snack	Hotel	Phone	Taxi	Cleaning	Tips	Misc.	TOTAL

Trip Seq. No.________ Page Total________

Crew Members________ Amount Forward________

Monthly Total________

Date	Flight No.	A/C Type	A/C ID	From	To	Time			Flight Conditions				
						Dep	Arr	TOTAL	Inst	App	Night		

Duty Start ________ Int'l Pay ________ Page Total ____ ____ ____ ____ ____ ____

Duty Stop ________ Night Pay ________ Amt Fwd ____ ____ ____ ____ ____ ____

Total Duty ________ Pay ________ Monthly Total ____ ____ ____ ____ ____ ____

Remarks: ____

____

Expenses Hours ________

Breakfast	Lunch	Dinner	Snack	Hotel	Phone	Taxi	Cleaning	Tips	Misc.	TOTAL

Trip Seq. No. ________________ Page Total ________

Crew Members ________________ Amount Forward ________

________________ Monthly Total ________

Date	Flight No.	A/C Type	A/C ID	From	To	Time			Flight Conditions				
						Dep	Arr	TOTAL	Inst	App	Night		

Duty Start________ Int'l Pay________ Page Total________

Duty Stop________ Night Pay________ Amt Fwd________

Total Duty________ Pay________ Monthly Total________

Remarks:________

Expenses Hours________

Breakfast	Lunch	Dinner	Snack	Hotel	Phone	Taxi	Cleaning	Tips	Misc.	TOTAL

Trip Seq. No.________ Page Total________

Crew Members________ Amount Forward________

Monthly Total________

Date	Flight No.	A/C Type	A/C ID	From	To	Time			Flight Conditions				
						Dep	Arr	TOTAL	Inst	App	Night		

Duty Start________ Int'l Pay________ Page Total____ ____ ____ ____ ____ ____

Duty Stop________ Night Pay________ Amt Fwd____ ____ ____ ____ ____ ____

Total Duty________ Pay________ Monthly Total____ ____ ____ ____ ____ ____

Remarks:________________________

Expenses Hours________

Breakfast	Lunch	Dinner	Snack	Hotel	Phone	Taxi	Cleaning	Tips	Misc.	TOTAL

Trip Seq. No.________ Page Total________

Crew Members________ Amount Forward________

Monthly Total________

Date	Flight No.	A/C Type	A/C ID	From	To	Time			Flight Conditions				
						Dep	Arr	TOTAL	Inst	App	Night		

Duty Start ________ Int'l Pay ________ Page Total ____ ____ ____ ____ ____ ____

Duty Stop ________ Night Pay ________ Amt Fwd ____ ____ ____ ____ ____ ____

Total Duty ________ Pay ________ Monthly Total ____ ____ ____ ____ ____ ____

Remarks: ________________________________

Expenses Hours ________

Breakfast	Lunch	Dinner	Snack	Hotel	Phone	Taxi	Cleaning	Tips	Misc.	TOTAL

Trip Seq. No. ____________ Page Total ________

Crew Members ________________________ Amount Forward ________

Monthly Total ________

Date	Flight No.	A/C Type	A/C ID	From	To	Time			Flight Conditions				
						Dep	Arr	TOTAL	Inst	App	Night		

Duty Start________ Int'l Pay________ Page Total____ ____ ____ ____ ____ ____

Duty Stop________ Night Pay________ Amt Fwd____ ____ ____ ____ ____ ____

Total Duty________ Pay________ Monthly Total____ ____ ____ ____ ____ ____

Remarks:________________________________

Expenses Hours________

Breakfast	Lunch	Dinner	Snack	Hotel	Phone	Taxi	Cleaning	Tips	Misc.	TOTAL

Trip Seq. No.____________ Page Total________

Crew Members____________________ Amount Forward________

Monthly Total________

Date	Flight No.	A/C Type	A/C ID	From	To	Time			Flight Conditions				
						Dep	Arr	TOTAL	Inst	App	Night		

Duty Start______ Int'l Pay______ Page Total______

Duty Stop______ Night Pay______ Amt Fwd______

Total Duty______ Pay______ Monthly Total______

Remarks:______

Expenses Hours______

Breakfast	Lunch	Dinner	Snack	Hotel	Phone	Taxi	Cleaning	Tips	Misc.	TOTAL

Trip Seq. No.______ Page Total______

Crew Members______ Amount Forward______

Monthly Total______

Date	Flight No.	A/C Type	A/C ID	From	To	Time			Flight Conditions				
						Dep	Arr	TOTAL	Inst	App	Night		

Duty Start________ Int'l Pay________ Page Total____ ____ ____ ____ ____ ____

Duty Stop________ Night Pay________ Amt Fwd____ ____ ____ ____ ____ ____

Total Duty________ Pay________ Monthly Total____ ____ ____ ____ ____ ____

Remarks:________________________

Expenses Hours________

Breakfast	Lunch	Dinner	Snack	Hotel	Phone	Taxi	Cleaning	Tips	Misc.	TOTAL

Trip Seq. No.________ Page Total________

Crew Members________ Amount Forward________

Monthly Total________

Date	Flight No.	A/C Type	A/C ID	From	To	Time			Flight Conditions				
						Dep	Arr	TOTAL	Inst	App	Night		

Duty Start ________ Int'l Pay ________ Page Total ____ ____ ____ ____ ____ ____

Duty Stop ________ Night Pay ________ Amt Fwd ____ ____ ____ ____ ____ ____

Total Duty ________ Pay ________ Monthly Total ____ ____ ____ ____ ____ ____

Remarks: ____________________

Expenses Hours ________

Breakfast	Lunch	Dinner	Snack	Hotel	Phone	Taxi	Cleaning	Tips	Misc.	TOTAL

Trip Seq. No. ________ Page Total ________

Crew Members ________ Amount Forward ________

Monthly Total ________

Date	Flight No.	A/C Type	A/C ID	From	To	Time			Flight Conditions				
						Dep	Arr	TOTAL	Inst	App	Night		

Duty Start________ Int'l Pay________ Page Total____ ____ ____ ____ ____ ____

Duty Stop________ Night Pay________ Amt Fwd____ ____ ____ ____ ____ ____

Total Duty________ Pay________ Monthly Total____ ____ ____ ____ ____ ____

Remarks:____

Expenses Hours________

Breakfast	Lunch	Dinner	Snack	Hotel	Phone	Taxi	Cleaning	Tips	Misc.	TOTAL

Trip Seq. No.________________ Page Total________

Crew Members________________________ Amount Forward________

Monthly Total________

Date	Flight No.	A/C Type	A/C ID	From	To	Time			Flight Conditions				
						Dep	Arr	TOTAL	Inst	App	Night		

Duty Start ____________ Int'l Pay ____________ Page Total ______ ______ ______ ______ ______ ______

Duty Stop ____________ Night Pay ____________ Amt Fwd ______ ______ ______ ______ ______ ______

Total Duty ____________ Pay ____________ Monthly Total ______ ______ ______ ______ ______ ______

Remarks: ____

____

Expenses Hours ____________

Breakfast	Lunch	Dinner	Snack	Hotel	Phone	Taxi	Cleaning	Tips	Misc.	TOTAL

Trip Seq. No. ____________________ Page Total ____________

Crew Members ____________________ Amount Forward ____________

____________________ Monthly Total ____________

Date	Flight No.	A/C Type	A/C ID	From	To	Time			Flight Conditions				
						Dep	Arr	TOTAL	Inst	App	Night		

Duty Start______ Int'l Pay______ Page Total______

Duty Stop______ Night Pay______ Amt Fwd______

Total Duty______ Pay______ Monthly Total______

Remarks:______

Expenses Hours______

Breakfast	Lunch	Dinner	Snack	Hotel	Phone	Taxi	Cleaning	Tips	Misc.	TOTAL

Trip Seq. No.______ Page Total______

Crew Members______ Amount Forward______

Monthly Total______

Date	Flight No.	A/C Type	A/C ID	From	To	Time			Flight Conditions				
						Dep	Arr	TOTAL	Inst	App	Night		

Duty Start ____________ Int'l Pay ____________ Page Total ______ ______ ______ ______ ______ ______

Duty Stop ____________ Night Pay ____________ Amt Fwd ______ ______ ______ ______ ______ ______

Total Duty ____________ Pay ____________ Monthly Total ______ ______ ______ ______ ______ ______

Remarks: ____

____

Expenses Hours ____________

Breakfast	Lunch	Dinner	Snack	Hotel	Phone	Taxi	Cleaning	Tips	Misc.	TOTAL

Trip Seq. No. ____________________ Page Total ____________

Crew Members ____________________ Amount Forward ____________

____________________ Monthly Total ____________

Date	Flight No.	A/C Type	A/C ID	From	To	Time			Flight Conditions				
						Dep	Arr	TOTAL	Inst	App	Night		

Duty Start______ Int'l Pay______ Page Total______

Duty Stop______ Night Pay______ Amt Fwd______

Total Duty______ Pay______ Monthly Total______

Remarks:______

Expenses Hours______

Breakfast	Lunch	Dinner	Snack	Hotel	Phone	Taxi	Cleaning	Tips	Misc.	TOTAL

Trip Seq. No.______ Page Total______

Crew Members______ Amount Forward______

Monthly Total______

Date	Flight No.	A/C Type	A/C ID	From	To	Time			Flight Conditions				
						Dep	Arr	TOTAL	Inst	App	Night		

Duty Start________ Int'l Pay________ Page Total____ ____ ____ ____ ____ ____

Duty Stop________ Night Pay________ Amt Fwd____ ____ ____ ____ ____ ____

Total Duty________ Pay________ Monthly Total____ ____ ____ ____ ____ ____

Remarks:________________________________

Expenses Hours________

Breakfast	Lunch	Dinner	Snack	Hotel	Phone	Taxi	Cleaning	Tips	Misc.	TOTAL

Trip Seq. No.________________ Page Total________

Crew Members________________ Amount Forward________

Monthly Total________

Date	Flight No.	A/C Type	A/C ID	From	To	Time			Flight Conditions				
						Dep	Arr	TOTAL	Inst	App	Night		

Duty Start______ Int'l Pay______ Page Total______ ______ ______ ______ ______ ______

Duty Stop______ Night Pay______ Amt Fwd______ ______ ______ ______ ______ ______

Total Duty______ Pay______ Monthly Total______ ______ ______ ______ ______ ______

Remarks:______

Expenses Hours______

Breakfast	Lunch	Dinner	Snack	Hotel	Phone	Taxi	Cleaning	Tips	Misc.	TOTAL

Trip Seq. No.______ Page Total______

Crew Members______ Amount Forward______

Monthly Total______

Date	Flight No.	A/C Type	A/C ID	From	To	Time			Flight Conditions				
						Dep	Arr	TOTAL	Inst	App	Night		

Duty Start________ Int'l Pay________ Page Total____ ____ ____ ____ ____ ____

Duty Stop________ Night Pay________ Amt Fwd____ ____ ____ ____ ____ ____

Total Duty________ Pay________ Monthly Total____ ____ ____ ____ ____ ____

Remarks:________________________________

Expenses Hours________

Breakfast	Lunch	Dinner	Snack	Hotel	Phone	Taxi	Cleaning	Tips	Misc.	TOTAL

Trip Seq. No.____________ Page Total________

Crew Members____________________ Amount Forward________

Monthly Total________

Date	Flight No.	A/C Type	A/C ID	From	To	Time			Flight Conditions				
						Dep	Arr	TOTAL	Inst	App	Night		

Duty Start ________ Int'l Pay ________ Page Total ____ ____ ____ ____ ____ ____

Duty Stop ________ Night Pay ________ Amt Fwd ____ ____ ____ ____ ____ ____

Total Duty ________ Pay ________ Monthly Total ____ ____ ____ ____ ____ ____

Remarks: ________________________

Expenses Hours ________

Breakfast	Lunch	Dinner	Snack	Hotel	Phone	Taxi	Cleaning	Tips	Misc.	TOTAL

Trip Seq. No. ________ Page Total ________

Crew Members ________ Amount Forward ________

Monthly Total ________

Date	Flight No.	A/C Type	A/C ID	From	To	Time			Flight Conditions				
						Dep	Arr	TOTAL	Inst	App	Night		

Duty Start ______ Int'l Pay ______ Page Total ______ ______ ______ ______ ______ ______

Duty Stop ______ Night Pay ______ Amt Fwd ______ ______ ______ ______ ______ ______

Total Duty ______ Pay ______ Monthly Total ______ ______ ______ ______ ______ ______

Remarks: ______

Expenses Hours ______

Breakfast	Lunch	Dinner	Snack	Hotel	Phone	Taxi	Cleaning	Tips	Misc.	TOTAL

Trip Seq. No. ______

Crew Members ______

Page Total ______

Amount Forward ______

Monthly Total ______

Date	Flight No.	A/C Type	A/C ID	From	To	Time			Flight Conditions				
						Dep	Arr	TOTAL	Inst	App	Night		

Duty Start________ Int'l Pay________ Page Total________

Duty Stop________ Night Pay________ Amt Fwd________

Total Duty________ Pay________ Monthly Total________

Remarks:________

Expenses Hours________

Breakfast	Lunch	Dinner	Snack	Hotel	Phone	Taxi	Cleaning	Tips	Misc.	TOTAL

Trip Seq. No.________ Page Total________

Crew Members________ Amount Forward________

Monthly Total________

Date	Flight No.	A/C Type	A/C ID	From	To	Time			Flight Conditions				
						Dep	Arr	TOTAL	Inst	App	Night		

Duty Start ______ Int'l Pay ______ Page Total ______ ______ ______ ______ ______ ______

Duty Stop ______ Night Pay ______ Amt Fwd ______ ______ ______ ______ ______ ______

Total Duty ______ Pay ______ Monthly Total ______ ______ ______ ______ ______ ______

Remarks: ______

Expenses Hours ______

Breakfast	Lunch	Dinner	Snack	Hotel	Phone	Taxi	Cleaning	Tips	Misc.	TOTAL

Trip Seq. No. ______ Page Total ______

Crew Members ______ Amount Forward ______

Monthly Total ______

Date	Flight No.	A/C Type	A/C ID	From	To	Time			Flight Conditions				
						Dep	Arr	TOTAL	Inst	App	Night		

Duty Start________ Int'l Pay________ Page Total____ ____ ____ ____ ____ ____

Duty Stop________ Night Pay________ Amt Fwd____ ____ ____ ____ ____ ____

Total Duty________ Pay________ Monthly Total____ ____ ____ ____ ____ ____

Remarks:________________________________

Expenses Hours________

Breakfast	Lunch	Dinner	Snack	Hotel	Phone	Taxi	Cleaning	Tips	Misc.	TOTAL

Trip Seq. No.____________ Page Total________

Crew Members____________ Amount Forward________

Monthly Total________

Date	Flight No.	A/C Type	A/C ID	From	To	Time			Flight Conditions				
						Dep	Arr	TOTAL	Inst	App	Night		

Duty Start________ Int'l Pay________ Page Total______ ______ ______ ______ ______ ______

Duty Stop________ Night Pay________ Amt Fwd______ ______ ______ ______ ______ ______

Total Duty________ Pay________ Monthly Total______ ______ ______ ______ ______ ______

Remarks:____

Expenses Hours________

Breakfast	Lunch	Dinner	Snack	Hotel	Phone	Taxi	Cleaning	Tips	Misc.	TOTAL

Trip Seq. No.____________ Page Total________

Crew Members________________________ Amount Forward________

Monthly Total________

Date	Flight No.	A/C Type	A/C ID	From	To	Time			Flight Conditions				
						Dep	Arr	TOTAL	Inst	App	Night		

Duty Start________ Int'l Pay________ Page Total____ ____ ____ ____ ____ ____

Duty Stop________ Night Pay________ Amt Fwd____ ____ ____ ____ ____ ____

Total Duty________ Pay________ Monthly Total____ ____ ____ ____ ____ ____

Remarks:________________________________

Expenses Hours________

Breakfast	Lunch	Dinner	Snack	Hotel	Phone	Taxi	Cleaning	Tips	Misc.	TOTAL

Trip Seq. No.____________ Page Total________

Crew Members____________ Amount Forward________

Monthly Total________

Date	Flight No.	A/C Type	A/C ID	From	To	Time			Flight Conditions				
						Dep	Arr	TOTAL	Inst	App	Night		

Duty Start________ Int'l Pay________ Page Total____ ____ ____ ____ ____ ____

Duty Stop________ Night Pay________ Amt Fwd____ ____ ____ ____ ____ ____

Total Duty________ Pay________ Monthly Total____ ____ ____ ____ ____ ____

Remarks:____

Expenses Hours________

Breakfast	Lunch	Dinner	Snack	Hotel	Phone	Taxi	Cleaning	Tips	Misc.	TOTAL

Trip Seq. No.____________ Page Total________

Crew Members________________________ Amount Forward________

Monthly Total________

Date	Flight No.	A/C Type	A/C ID	From	To	Time			Flight Conditions				
						Dep	Arr	TOTAL	Inst	App	Night		

Duty Start________ Int'l Pay________ Page Total____ ____ ____ ____ ____ ____

Duty Stop________ Night Pay________ Amt Fwd____ ____ ____ ____ ____ ____

Total Duty________ Pay________ Monthly Total____ ____ ____ ____ ____ ____

Remarks:________________________________

____

Expenses Hours________

Breakfast	Lunch	Dinner	Snack	Hotel	Phone	Taxi	Cleaning	Tips	Misc.	TOTAL

Trip Seq. No.________________ Page Total________

Crew Members________________________ Amount Forward________

________________________________ Monthly Total________

Date	Flight No.	A/C Type	A/C ID	From	To	Time			Flight Conditions				
						Dep	Arr	TOTAL	Inst	App	Night		

Duty Start ____________ Int'l Pay ____________ Page Total ______ ______ ______ ______ ______ ______

Duty Stop ____________ Night Pay ____________ Amt Fwd ______ ______ ______ ______ ______ ______

Total Duty ____________ Pay ____________ Monthly Total ______ ______ ______ ______ ______ ______

Remarks: ____

____

Expenses Hours ____________

Breakfast	Lunch	Dinner	Snack	Hotel	Phone	Taxi	Cleaning	Tips	Misc.	TOTAL

Trip Seq. No. ____________________ Page Total ____________

Crew Members ____________________ Amount Forward ____________

____________________ Monthly Total ____________

Date	Flight No.	A/C Type	A/C ID	From	To	Time			Flight Conditions				
						Dep	Arr	TOTAL	Inst	App	Night		

Duty Start________ Int'l Pay________ Page Total____ ____ ____ ____ ____ ____

Duty Stop________ Night Pay________ Amt Fwd____ ____ ____ ____ ____ ____

Total Duty________ Pay________ Monthly Total____ ____ ____ ____ ____ ____

Remarks:________________________

Expenses Hours________

Breakfast	Lunch	Dinner	Snack	Hotel	Phone	Tax	Cleaning	Tips	Misc.	TOTAL

Trip Seq. No.____________ Page Total________

Crew Members____________ Amount Forward________

Monthly Total________

Date	Flight No.	A/C Type	A/C ID	From	To	Time			Flight Conditions				
						Dep	Arr	TOTAL	Inst	App	Night		

Duty Start________ Int'l Pay________ Page Total____ ____ ____ ____ ____ ____

Duty Stop________ Night Pay________ Amt Fwd____ ____ ____ ____ ____ ____

Total Duty________ Pay________ Monthly Total____ ____ ____ ____ ____ ____

Remarks:________________________

Expenses Hours________

Breakfast	Lunch	Dinner	Snack	Hotel	Phone	Taxi	Cleaning	Tips	Misc.	TOTAL

Trip Seq. No.________ Page Total________

Crew Members________________ Amount Forward________

Monthly Total________

Date	Flight No.	A/C Type	A/C ID	From	To	Time			Flight Conditions				
						Dep	Arr	TOTAL	Inst	App	Night		

Duty Start________ Int'l Pay________ Page Total____ ____ ____ ____ ____ ____

Duty Stop________ Night Pay________ Amt Fwd____ ____ ____ ____ ____ ____

Total Duty________ Pay________ Monthly Total____ ____ ____ ____ ____ ____

Remarks:________________________________

Expenses Hours________

Breakfast	Lunch	Dinner	Snack	Hotel	Phone	Taxi	Cleaning	Tips	Misc.	TOTAL

Trip Seq. No.________ Page Total________

Crew Members________________ Amount Forward________

Monthly Total________

Date	Flight No.	A/C Type	A/C ID	From	To	Time			Flight Conditions				
						Dep	Arr	TOTAL	Inst	App	Night		

Duty Start________ Int'l Pay________ Page Total____ ____ ____ ____ ____ ____

Duty Stop________ Night Pay________ Amt Fwd____ ____ ____ ____ ____ ____

Total Duty________ Pay________ Monthly Total____ ____ ____ ____ ____ ____

Remarks:________________________________

Expenses Hours________

Breakfast	Lunch	Dinner	Snack	Hotel	Phone	Taxi	Cleaning	Tips	Misc.	TOTAL

Trip Seq. No.____________ Page Total________

Crew Members____________________ Amount Forward________

Monthly Total________

Date	Flight No.	A/C Type	A/C ID	From	To	Time			Flight Conditions				
						Dep	Arr	TOTAL	Inst	App	Night		

Duty Start ______ Int'l Pay ______ Page Total ______ ______ ______ ______ ______ ______

Duty Stop ______ Night Pay ______ Amt Fwd ______ ______ ______ ______ ______ ______

Total Duty ______ Pay ______ Monthly Total ______ ______ ______ ______ ______ ______

Remarks: ______

Expenses Hours ______

Breakfast	Lunch	Dinner	Snack	Hotel	Phone	Taxi	Cleaning	Tips	Misc.	TOTAL

Trip Seq. No. ______

Crew Members ______

Page Total ______

Amount Forward ______

Monthly Total ______

Date	Flight No.	A/C Type	A/C ID	From	To	Time			Flight Conditions				
						Dep	Arr	TOTAL	Inst	App	Night		

Duty Start ______ Int'l Pay ______ Page Total ______ ______ ______ ______ ______ ______

Duty Stop ______ Night Pay ______ Amt Fwd ______ ______ ______ ______ ______ ______

Total Duty ______ Pay ______ Monthly Total ______ ______ ______ ______ ______ ______

Remarks: ______

______

Expenses Hours ______

Breakfast	Lunch	Dinner	Snack	Hotel	Phone	Taxi	Cleaning	Tips	Misc.	TOTAL

Trip Seq. No. ______ Page Total ______

Crew Members ______ Amount Forward ______

______ Monthly Total ______

Date	Flight No.	A/C Type	A/C ID	From	To	Time			Flight Conditions				
						Dep	Arr	TOTAL	Inst	App	Night		

Duty Start________ Int'l Pay________ Page Total____ ____ ____ ____ ____ ____

Duty Stop________ Night Pay________ Amt Fwd____ ____ ____ ____ ____ ____

Total Duty________ Pay________ Monthly Total____ ____ ____ ____ ____ ____

Remarks:____

Expenses Hours________

Breakfast	Lunch	Dinner	Snack	Hotel	Phone	Tax	Cleaning	Tips	Misc.	TOTAL

Trip Seq. No.________________ Page Total________

Crew Members________________________ Amount Forward________

Monthly Total________

Date	Flight No.	A/C Type	A/C ID	From	To	Time			Flight Conditions				
						Dep	Arr	TOTAL	Inst	App	Night		

Duty Start ________ Int'l Pay ________ Page Total ____ ____ ____ ____ ____ ____

Duty Stop ________ Night Pay ________ Amt Fwd ____ ____ ____ ____ ____ ____

Total Duty ________ Pay ________ Monthly Total ____ ____ ____ ____ ____ ____

Remarks: ________________________________

Expenses Hours ________

Breakfast	Lunch	Dinner	Snack	Hotel	Phone	Taxi	Cleaning	Tips	Misc.	TOTAL

Trip Seq. No. ____________ Page Total ________

Crew Members ________________________ Amount Forward ________

Monthly Total ________

Date	Flight No.	A/C Type	A/C ID	From	To	Time			Flight Conditions				
						Dep	Arr	TOTAL	Inst	App	Night		

Duty Start_______ Int'l Pay_______ Page Total____ ____ ____ ____ ____ ____

Duty Stop_______ Night Pay_______ Amt Fwd____ ____ ____ ____ ____ ____

Total Duty_______ Pay_______ Monthly Total____ ____ ____ ____ ____ ____

Remarks:______________________________

Expenses Hours_______

Breakfast	Lunch	Dinner	Snack	Hotel	Phone	Taxi	Cleaning	Tips	Misc.	TOTAL

Trip Seq. No.__________ Page Total_______

Crew Members____________________ Amount Forward_______

Monthly Total_______

Date	Flight No.	A/C Type	A/C ID	From	To	Time			Flight Conditions				
						Dep	Arr	TOTAL	Inst	App	Night		

Duty Start__________ Int'l Pay__________ Page Total____ ____ ____ ____ ____ ____

Duty Stop__________ Night Pay__________ Amt Fwd____ ____ ____ ____ ____ ____

Total Duty__________ Pay__________ Monthly Total____ ____ ____ ____ ____ ____

Remarks:____

Expenses Hours__________

Breakfast	Lunch	Dinner	Snack	Hotel	Phone	Taxi	Cleaning	Tips	Misc.	TOTAL

Trip Seq. No.________________ Page Total__________

Crew Members____________________________ Amount Forward__________

Monthly Total__________

Date	Flight No.	A/C Type	A/C ID	From	To	Time			Flight Conditions				
						Dep	Arr	TOTAL	Inst	App	Night		

Duty Start________ Int'l Pay________ Page Total____ ____ ____ ____ ____ ____

Duty Stop________ Night Pay________ Amt Fwd____ ____ ____ ____ ____ ____

Total Duty________ Pay________ Monthly Total____ ____ ____ ____ ____ ____

Remarks:________________________

Expenses Hours________

Breakfast	Lunch	Dinner	Snack	Hotel	Phone	Taxi	Cleaning	Tips	Misc.	TOTAL

Trip Seq. No.____________ Page Total________

Crew Members____________________ Amount Forward________

Monthly Total________

Date	Flight No.	A/C Type	A/C ID	From	To	Time			Flight Conditions				
						Dep	Arr	TOTAL	Inst	App	Night		

Duty Start ______ Int'l Pay ______ Page Total ______ ______ ______ ______ ______ ______

Duty Stop ______ Night Pay ______ Amt Fwd ______ ______ ______ ______ ______ ______

Total Duty ______ Pay ______ Monthly Total ______ ______ ______ ______ ______ ______

Remarks: ______

Expenses Hours ______

Breakfast	Lunch	Dinner	Snack	Hotel	Phone	Taxi	Cleaning	Tips	Misc.	TOTAL

Trip Seq. No. ______ Page Total ______

Crew Members ______ Amount Forward ______

Monthly Total ______

Date	Flight No.	A/C Type	A/C ID	From	To	Time			Flight Conditions				
						Dep	Arr	TOTAL	Inst	App	Night		

Duty Start________ Int'l Pay________ Page Total________

Duty Stop________ Night Pay________ Amt Fwd________

Total Duty________ Pay________ Monthly Total________

Remarks:________

Expenses Hours________

Breakfast	Lunch	Dinner	Snack	Hotel	Phone	Taxi	Cleaning	Tips	Misc.	TOTAL

Trip Seq. No.________ Page Total________

Crew Members________ Amount Forward________

Monthly Total________

Date	Flight No.	A/C Type	A/C ID	From	To	Time			Flight Conditions				
						Dep	Arr	TOTAL	Inst	App	Night		

Duty Start________ Int'l Pay________ Page Total____ ____ ____ ____ ____ ____

Duty Stop________ Night Pay________ Amt Fwd____ ____ ____ ____ ____ ____

Total Duty________ Pay________ Monthly Total____ ____ ____ ____ ____ ____

Remarks:________________________

Expenses Hours________

Breakfast	Lunch	Dinner	Snack	Hotel	Phone	Taxi	Cleaning	Tips	Misc.	TOTAL

Trip Seq. No.________ Page Total________

Crew Members________ Amount Forward________

Monthly Total________

Date	Flight No.	A/C Type	A/C ID	From	To	Time			Flight Conditions				
						Dep	Arr	TOTAL	Inst	App	Night		

Duty Start________ Int'l Pay________ Page Total____ ____ ____ ____ ____ ____

Duty Stop________ Night Pay________ Amt Fwd____ ____ ____ ____ ____ ____

Total Duty________ Pay________ Monthly Total____ ____ ____ ____ ____ ____

Remarks:________________________________

Expenses Hours________

Breakfast	Lunch	Dinner	Snack	Hotel	Phone	Taxi	Cleaning	Tips	Misc.	TOTAL

Trip Seq. No.____________ Page Total________

Crew Members____________ Amount Forward________

Monthly Total________

Date	Flight No.	A/C Type	A/C ID	From	To	Time			Flight Conditions				
						Dep	Arr	TOTAL	Inst	App	Night		

Duty Start ______ Int'l Pay ______ Page Total ______ ______ ______ ______ ______ ______

Duty Stop ______ Night Pay ______ Amt Fwd ______ ______ ______ ______ ______ ______

Total Duty ______ Pay ______ Monthly Total ______ ______ ______ ______ ______ ______

Remarks: ______

______

Expenses Hours ______

Breakfast	Lunch	Dinner	Snack	Hotel	Phone	Taxi	Cleaning	Tips	Misc.	TOTAL

Trip Seq. No. ______ Page Total ______

Crew Members ______ Amount Forward ______

______ Monthly Total ______

Date	Flight No.	A/C Type	A/C ID	From	To	Time			Flight Conditions				
						Dep	Arr	TOTAL	Inst	App	Night		

Duty Start________ Int'l Pay________ Page Total____ ____ ____ ____ ____ ____

Duty Stop________ Night Pay________ Amt Fwd____ ____ ____ ____ ____ ____

Total Duty________ Pay________ Monthly Total____ ____ ____ ____ ____ ____

Remarks:________________________________

____

Expenses Hours________

Breakfast	Lunch	Dinner	Snack	Hotel	Phone	Taxi	Cleaning	Tips	Misc.	TOTAL

Trip Seq. No.________________ Page Total________

Crew Members________________________ Amount Forward________

____________________________ Monthly Total________

Date	Flight No.	A/C Type	A/C ID	From	To	Time			Flight Conditions				
						Dep	Arr	TOTAL	Inst	App	Night		

Duty Start______ Int'l Pay______ Page Total______

Duty Stop______ Night Pay______ Amt Fwd______

Total Duty______ Pay______ Monthly Total______

Remarks:______

Expenses Hours______

Breakfast	Lunch	Dinner	Snack	Hotel	Phone	Taxi	Cleaning	Tips	Misc.	TOTAL

Trip Seq. No.______ Page Total______

Crew Members______ Amount Forward______

Monthly Total______

Date	Flight No.	A/C Type	A/C ID	From	To	Time			Flight Conditions				
						Dep	Arr	TOTAL	Inst	App	Night		

Duty Start______________ Int'l Pay______________ Page Total______ ______ ______ ______ ______ ______

Duty Stop______________ Night Pay______________ Amt Fwd______ ______ ______ ______ ______ ______

Total Duty______________ Pay______________ Monthly Total______ ______ ______ ______ ______ ______

Remarks:____

____

Expenses Hours______________

Breakfast	Lunch	Dinner	Snack	Hotel	Phone	Taxi	Cleaning	Tips	Misc.	TOTAL

Trip Seq. No.______________________ Page Total______________

Crew Members____ Amount Forward______________

____ Monthly Total______________

Date	Flight No.	A/C Type	A/C ID	From	To	Time			Flight Conditions				
						Dep	Arr	TOTAL	Inst	App	Night		

Duty Start________ Int'l Pay________ Page Total____ ____ ____ ____ ____ ____

Duty Stop________ Night Pay________ Amt Fwd____ ____ ____ ____ ____ ____

Total Duty________ Pay________ Monthly Total____ ____ ____ ____ ____ ____

Remarks:________________________

Expenses Hours________

Breakfast	Lunch	Dinner	Snack	Hotel	Phone	Taxi	Cleaning	Tips	Misc.	TOTAL

Trip Seq. No.________ Page Total________

Crew Members________ Amount Forward________

Monthly Total________

Date	Flight No.	A/C Type	A/C ID	From	To	Time			Flight Conditions				
						Dep	Arr	TOTAL	Inst	App	Night		

Duty Start________ Int'l Pay________ Page Total______

Duty Stop________ Night Pay________ Amt Fwd______

Total Duty________ Pay________ Monthly Total______

Remarks:________

Expenses Hours________

Breakfast	Lunch	Dinner	Snack	Hotel	Phone	Taxi	Cleaning	Tips	Misc.	TOTAL

Trip Seq. No.________ Page Total______

Crew Members________ Amount Forward______

Monthly Total______

Date	Flight No.	A/C Type	A/C ID	From	To	Time			Flight Conditions				
						Dep	Arr	TOTAL	Inst	App	Night		

Duty Start______ Int'l Pay______ Page Total______

Duty Stop______ Night Pay______ Amt Fwd______

Total Duty______ Pay______ Monthly Total______

Remarks:______

Expenses Hours______

Breakfast	Lunch	Dinner	Snack	Hotel	Phone	Taxi	Cleaning	Tips	Misc.	TOTAL

Trip Seq. No.______ Page Total______

Crew Members______ Amount Forward______

Monthly Total______

Date	Flight No.	A/C Type	A/C ID	From	To	Time			Flight Conditions				
						Dep	Arr	TOTAL	Inst	App	Night		

Duty Start ____________ Int'l Pay ____________ Page Total ______ ______ ______ ______ ______ ______

Duty Stop ____________ Night Pay ____________ Amt Fwd ______ ______ ______ ______ ______ ______

Total Duty ____________ Pay ____________ Monthly Total ______ ______ ______ ______ ______ ______

Remarks: ____

____

Expenses Hours ____________

Breakfast	Lunch	Dinner	Snack	Hotel	Phone	Taxi	Cleaning	Tips	Misc.	TOTAL

Trip Seq. No. ____________________ Page Total ____________

Crew Members ____________________ Amount Forward ____________

____________________ Monthly Total ____________

Date	Flight No.	A/C Type	A/C ID	From	To	Time			Flight Conditions				
						Dep	Arr	TOTAL	Inst	App	Night		

Duty Start________ Int'l Pay________ Page Total____ ____ ____ ____ ____ ____

Duty Stop________ Night Pay________ Amt Fwd____ ____ ____ ____ ____ ____

Total Duty________ Pay________ Monthly Total____ ____ ____ ____ ____ ____

Remarks:________________________

Expenses Hours________

Breakfast	Lunch	Dinner	Snack	Hotel	Phone	Taxi	Cleaning	Tips	Misc.	TOTAL

Trip Seq. No.________ Page Total________

Crew Members________ Amount Forward________

Monthly Total________

Date	Flight No.	A/C Type	A/C ID	From	To	Time			Flight Conditions				
						Dep	Arr	TOTAL	Inst	App	Night		

Duty Start______ Int'l Pay______ Page Total______

Duty Stop______ Night Pay______ Amt Fwd______

Total Duty______ Pay______ Monthly Total______

Remarks:______

Expenses Hours______

Breakfast	Lunch	Dinner	Snack	Hotel	Phone	Taxi	Cleaning	Tips	Misc.	TOTAL

Trip Seq. No.______ Page Total______

Crew Members______ Amount Forward______

Monthly Total______

Date	Flight No.	A/C Type	A/C ID	From	To	Time			Flight Conditions				
						Dep	Arr	TOTAL	Inst	App	Night		

Duty Start______________ Int'l Pay______________ Page Total______ ______ ______ ______ ______ ______

Duty Stop______________ Night Pay______________ Amt Fwd______ ______ ______ ______ ______ ______

Total Duty______________ Pay______________ Monthly Total______ ______ ______ ______ ______ ______

Remarks:____

____

Expenses Hours______________

Breakfast	Lunch	Dinner	Snack	Hotel	Phone	Taxi	Cleaning	Tips	Misc.	TOTAL

Trip Seq. No.______________________ Page Total______________

Crew Members____ Amount Forward______________

____ Monthly Total______________

Date	Flight No.	A/C Type	A/C ID	From	To	Time			Flight Conditions				
						Dep	Arr	TOTAL	Inst	App	Night		

Duty Start________ Int'l Pay________ Page Total____ ____ ____ ____ ____ ____

Duty Stop________ Night Pay________ Amt Fwd____ ____ ____ ____ ____ ____

Total Duty________ Pay________ Monthly Total____ ____ ____ ____ ____ ____

Remarks:________________

Expenses Hours________

Breakfast	Lunch	Dinner	Snack	Hotel	Phone	Taxi	Cleaning	Tips	Misc.	TOTAL

Trip Seq. No.________ Page Total________

Crew Members________ Amount Forward________

Monthly Total________

Date	Flight No.	A/C Type	A/C ID	From	To	Time			Flight Conditions				
						Dep	Arr	TOTAL	Inst	App	Night		

Duty Start________ Int'l Pay________ Page Total____ ____ ____ ____ ____ ____

Duty Stop________ Night Pay________ Amt Fwd____ ____ ____ ____ ____ ____

Total Duty________ Pay________ Monthly Total____ ____ ____ ____ ____ ____

Remarks:________________

Expenses Hours________

Breakfast	Lunch	Dinner	Snack	Hotel	Phone	Taxi	Cleaning	Tips	Misc.	TOTAL

Trip Seq. No.________ Page Total________

Crew Members________ Amount Forward________

Monthly Total________

Date	Flight No.	A/C Type	A/C ID	From	To	Time			Flight Conditions				
						Dep	Arr	TOTAL	Inst	App	Night		

Duty Start ____________ Int'l Pay ____________ Page Total ____ ____ ____ ____ ____ ____

Duty Stop ____________ Night Pay ____________ Amt Fwd ____ ____ ____ ____ ____ ____

Total Duty ____________ Pay ____________ Monthly Total ____ ____ ____ ____ ____ ____

Remarks: ____

____

Expenses Hours ____________

Breakfast	Lunch	Dinner	Snack	Hotel	Phone	Taxi	Cleaning	Tips	Misc.	TOTAL

Trip Seq. No. ____________________ Page Total ____________

Crew Members ______________________________ Amount Forward ____________

______________________________ Monthly Total ____________

Date	Flight No.	A/C Type	A/C ID	From	To	Time			Flight Conditions				
						Dep	Arr	TOTAL	Inst	App	Night		

Duty Start__________ Int'l Pay__________ Page Total_____ _____ _____ _____ _____ _____

Duty Stop__________ Night Pay__________ Amt Fwd_____ _____ _____ _____ _____ _____

Total Duty__________ Pay__________ Monthly Total_____ _____ _____ _____ _____ _____

Remarks:____

Expenses Hours__________

Breakfast	Lunch	Dinner	Snack	Hotel	Phone	Taxi	Cleaning	Tips	Misc.	TOTAL

Trip Seq. No.____________________ Page Total__________

Crew Members______________________________ Amount Forward__________

Monthly Total__________

Date	Flight No.	A/C Type	A/C ID	From	To	Time			Flight Conditions				
						Dep	Arr	TOTAL	Inst	App	Night		

Duty Start________ Int'l Pay________ Page Total____ ____ ____ ____ ____ ____

Duty Stop________ Night Pay________ Amt Fwd____ ____ ____ ____ ____ ____

Total Duty________ Pay________ Monthly Total____ ____ ____ ____ ____ ____

Remarks:________________________________

Expenses Hours________

Breakfast	Lunch	Dinner	Snack	Hotel	Phone	Taxi	Cleaning	Tips	Misc.	TOTAL

Trip Seq. No.____________ Page Total________

Crew Members____________________ Amount Forward________

Monthly Total________

Date	Flight No.	A/C Type	A/C ID	From	To	Time			Flight Conditions				
						Dep	Arr	TOTAL	Inst	App	Night		

Duty Start________ Int'l Pay________ Page Total____ ____ ____ ____ ____ ____

Duty Stop________ Night Pay________ Amt Fwd____ ____ ____ ____ ____ ____

Total Duty________ Pay________ Monthly Total____ ____ ____ ____ ____ ____

Remarks:________________________________

Expenses Hours________

Breakfast	Lunch	Dinner	Snack	Hotel	Phone	Taxi	Cleaning	Tips	Misc.	TOTAL

Trip Seq. No.____________ Page Total________

Crew Members____________________ Amount Forward________

______________________________ Monthly Total________

Date	Flight No.	A/C Type	A/C ID	From	To	Time			Flight Conditions				
						Dep	Arr	TOTAL	Inst	App	Night		

Duty Start______ Int'l Pay______ Page Total______

Duty Stop______ Night Pay______ Amt Fwd______

Total Duty______ Pay______ Monthly Total______

Remarks:______

Expenses Hours______

Breakfast	Lunch	Dinner	Snack	Hotel	Phone	Taxi	Cleaning	Tips	Misc.	TOTAL

Trip Seq. No.______ Page Total______

Crew Members______ Amount Forward______

Monthly Total______

Date	Flight No.	A/C Type	A/C ID	From	To	Time			Flight Conditions				
						Dep	Arr	TOTAL	Inst	App	Night		

Duty Start________ Int'l Pay________ Page Total________

Duty Stop________ Night Pay________ Amt Fwd________

Total Duty________ Pay________ Monthly Total________

Remarks:________

Expenses Hours________

Breakfast	Lunch	Dinner	Snack	Hotel	Phone	Taxi	Cleaning	Tips	Misc.	TOTAL

Trip Seq. No.________ Page Total________

Crew Members________ Amount Forward________

Monthly Total________

Date	Flight No.	A/C Type	A/C ID	From	To	Time			Flight Conditions				
						Dep	Arr	TOTAL	Inst	App	Night		

Duty Start________ Int'l Pay________ Page Total____ ____ ____ ____ ____ ____

Duty Stop________ Night Pay________ Amt Fwd____ ____ ____ ____ ____ ____

Total Duty________ Pay________ Monthly Total____ ____ ____ ____ ____ ____

Remarks:____

Expenses Hours________

Breakfast	Lunch	Dinner	Snack	Hotel	Phone	Taxi	Cleaning	Tips	Misc.	TOTAL

Trip Seq. No.________________ Page Total________

Crew Members____________________________ Amount Forward________

Monthly Total________

Date	Flight No.	A/C Type	A/C ID	From	To	Time			Flight Conditions				
						Dep	Arr	TOTAL	Inst	App	Night		

Duty Start______ Int'l Pay______ Page Total______

Duty Stop______ Night Pay______ Amt Fwd______

Total Duty______ Pay______ Monthly Total______

Remarks:______

Expenses Hours______

Breakfast	Lunch	Dinner	Snack	Hotel	Phone	Taxi	Cleaning	Tips	Misc.	TOTAL

Trip Seq. No.______ Page Total______

Crew Members______ Amount Forward______

Monthly Total______

Date	Flight No.	A/C Type	A/C ID	From	To	Time			Flight Conditions				
						Dep	Arr	TOTAL	Inst	App	Night		

Duty Start______ Int'l Pay______ Page Total______

Duty Stop______ Night Pay______ Amt Fwd______

Total Duty______ Pay______ Monthly Total______

Remarks:______

Expenses Hours______

Breakfast	Lunch	Dinner	Snack	Hotel	Phone	Taxi	Cleaning	Tips	Misc.	TOTAL

Trip Seq. No.______ Page Total______

Crew Members______ Amount Forward______

Monthly Total______

Date	Flight No.	A/C Type	A/C ID	From	To	Time			Flight Conditions				
						Dep	Arr	TOTAL	Inst	App	Night		

Duty Start________ Int'l Pay________ Page Total____ ____ ____ ____ ____ ____

Duty Stop________ Night Pay________ Amt Fwd____ ____ ____ ____ ____ ____

Total Duty________ Pay________ Monthly Total____ ____ ____ ____ ____ ____

Remarks:________________________

Expenses Hours________

Breakfast	Lunch	Dinner	Snack	Hotel	Phone	Taxi	Cleaning	Tips	Misc.	TOTAL

Trip Seq. No.____________ Page Total________

Crew Members____________ Amount Forward________

Monthly Total________

Date	Flight No.	A/C Type	A/C ID	From	To	Time			Flight Conditions				
						Dep	Arr	TOTAL	Inst	App	Night		

Duty Start______ Int'l Pay______ Page Total______

Duty Stop______ Night Pay______ Amt Fwd______

Total Duty______ Pay______ Monthly Total______

Remarks:______

Expenses Hours______

Breakfast	Lunch	Dinner	Snack	Hotel	Phone	Taxi	Cleaning	Tips	Misc.	TOTAL

Trip Seq. No.______ Page Total______

Crew Members______ Amount Forward______

Monthly Total______

Date	Flight No.	A/C Type	A/C ID	From	To	Time			Flight Conditions				
						Dep	Arr	TOTAL	Inst	App	Night		

Duty Start______ Int'l Pay______ Page Total______

Duty Stop______ Night Pay______ Amt Fwd______

Total Duty______ Pay______ Monthly Total______

Remarks:______

Expenses Hours______

Breakfast	Lunch	Dinner	Snack	Hotel	Phone	Taxi	Cleaning	Tips	Misc.	TOTAL

Trip Seq. No.______ Page Total______

Crew Members______ Amount Forward______

Monthly Total______

Date	Flight No.	A/C Type	A/C ID	From	To	Time			Flight Conditions				
						Dep	Arr	TOTAL	Inst	App	Night		

Duty Start ______ Int'l Pay ______ Page Total ______ ______ ______ ______ ______ ______

Duty Stop ______ Night Pay ______ Amt Fwd ______ ______ ______ ______ ______ ______

Total Duty ______ Pay ______ Monthly Total ______ ______ ______ ______ ______ ______

Remarks: ______

Expenses Hours ______

Breakfast	Lunch	Dinner	Snack	Hotel	Phone	Taxi	Cleaning	Tips	Misc.	TOTAL

Trip Seq. No. ______ Page Total ______

Crew Members ______ Amount Forward ______

Monthly Total ______

Date	Flight No.	A/C Type	A/C ID	From	To	Time			Flight Conditions				
						Dep	Arr	TOTAL	Inst	App	Night		

Duty Start________ Int'l Pay________ Page Total____ ____ ____ ____ ____ ____

Duty Stop________ Night Pay________ Amt Fwd____ ____ ____ ____ ____ ____

Total Duty________ Pay________ Monthly Total____ ____ ____ ____ ____ ____

Remarks:________________________

Expenses Hours________

Breakfast	Lunch	Dinner	Snack	Hotel	Phone	Taxi	Cleaning	Tips	Misc.	TOTAL

Trip Seq. No.________ Page Total________

Crew Members________________ Amount Forward________

Monthly Total________

Date	Flight No.	A/C Type	A/C ID	From	To	Time			Flight Conditions				
						Dep	Arr	TOTAL	Inst	App	Night		

Duty Start______ Int'l Pay______ Page Total______

Duty Stop______ Night Pay______ Amt Fwd______

Total Duty______ Pay______ Monthly Total______

Remarks:______

Expenses Hours______

Breakfast	Lunch	Dinner	Snack	Hotel	Phone	Taxi	Cleaning	Tips	Misc.	TOTAL

Trip Seq. No.______ Page Total______

Crew Members______ Amount Forward______

Monthly Total______

Date	Flight No.	A/C Type	A/C ID	From	To	Time			Flight Conditions				
						Dep	Arr	TOTAL	Inst	App	Night		

Duty Start________ Int'l Pay________ Page Total____ ____ ____ ____ ____ ____

Duty Stop________ Night Pay________ Amt Fwd____ ____ ____ ____ ____ ____

Total Duty________ Pay________ Monthly Total____ ____ ____ ____ ____ ____

Remarks:________________________________

Expenses Hours________

Breakfast	Lunch	Dinner	Snack	Hotel	Phone	Taxi	Cleaning	Tips	Misc.	TOTAL

Trip Seq. No.____________ Page Total________

Crew Members____________ Amount Forward________

Monthly Total________

Date	Flight No.	A/C Type	A/C ID	From	To	Time			Flight Conditions				
						Dep	Arr	TOTAL	Inst	App	Night		

Duty Start ______ Int'l Pay ______ Page Total ______ ______ ______ ______ ______ ______

Duty Stop ______ Night Pay ______ Amt Fwd ______ ______ ______ ______ ______ ______

Total Duty ______ Pay ______ Monthly Total ______ ______ ______ ______ ______ ______

Remarks: ______

Expenses Hours ______

Breakfast	Lunch	Dinner	Snack	Hotel	Phone	Taxi	Cleaning	Tips	Misc.	TOTAL

Trip Seq. No. ______ Page Total ______

Crew Members ______ Amount Forward ______

Monthly Total ______

Date	Flight No.	A/C Type	A/C ID	From	To	Time			Flight Conditions				
						Dep	Arr	TOTAL	Inst	App	Night		

Duty Start______ Int'l Pay______ Page Total______

Duty Stop______ Night Pay______ Amt Fwd______

Total Duty______ Pay______ Monthly Total______

Remarks:______

Expenses Hours______

Breakfast	Lunch	Dinner	Snack	Hotel	Phone	Taxi	Cleaning	Tips	Misc.	TOTAL

Trip Seq. No.______ Page Total______

Crew Members______ Amount Forward______

Monthly Total______

Date	Flight No.	A/C Type	A/C ID	From	To	Time			Flight Conditions				
						Dep	Arr	TOTAL	Inst	App	Night		

Duty Start ______ Int'l Pay ______ Page Total ______ ______ ______ ______ ______ ______

Duty Stop ______ Night Pay ______ Amt Fwd ______ ______ ______ ______ ______ ______

Total Duty ______ Pay ______ Monthly Total ______ ______ ______ ______ ______ ______

Remarks: ______

Expenses Hours ______

Breakfast	Lunch	Dinner	Snack	Hotel	Phone	Taxi	Cleaning	Tips	Misc.	TOTAL

Trip Seq. No. ______ Page Total ______

Crew Members ______ Amount Forward ______

Monthly Total ______

Date	Flight No.	A/C Type	A/C ID	From	To	Time			Flight Conditions				
						Dep	Arr	TOTAL	Inst	App	Night		

Duty Start ______ Int'l Pay ______ Page Total ______ ______ ______ ______ ______ ______

Duty Stop ______ Night Pay ______ Amt Fwd ______ ______ ______ ______ ______ ______

Total Duty ______ Pay ______ Monthly Total ______ ______ ______ ______ ______ ______

Remarks: ______

______

Expenses Hours ______

Breakfast	Lunch	Dinner	Snack	Hotel	Phone	Taxi	Cleaning	Tips	Misc.	TOTAL

Trip Seq. No. ______ Page Total ______

Crew Members ______ Amount Forward ______

______ Monthly Total ______

Date	Flight No.	A/C Type	A/C ID	From	To	Time			Flight Conditions				
						Dep	Arr	TOTAL	Inst	App	Night		

Duty Start________ Int'l Pay________ Page Total____ ____ ____ ____ ____ ____

Duty Stop________ Night Pay________ Amt Fwd____ ____ ____ ____ ____ ____

Total Duty________ Pay________ Monthly Total____ ____ ____ ____ ____ ____

Remarks:________________________________

Expenses Hours________

Breakfast	Lunch	Dinner	Snack	Hotel	Phone	Taxi	Cleaning	Tips	Misc.	TOTAL

Trip Seq. No.____________ Page Total________

Crew Members____________________ Amount Forward________

Monthly Total________

SUMMARY

YEAR 20 ____	Expenses				Time				
	Spent	Received	Hours	TOTAL	Inst	App	Night	Total	
Previous Book									
JANUARY									
FEBRUARY									
MARCH									
APRIL									
MAY									
JUNE									
JULY									
AUGUST									
SEPTEMBER									
OCTOBER									
NOVEMBER									
DECEMBER									
TOTAL									

SUMMARY

YEAR	Expenses				Time				
20 ____	Spent	Received	Hours	TOTAL	Inst	App	Night	Total	
Previous Book									
JANUARY									
FEBRUARY									
MARCH									
APRIL									
MAY									
JUNE									
JULY									
AUGUST									
SEPTEMBER									
OCTOBER									
NOVEMBER									
DECEMBER									
TOTAL									

Name	Telephone

Notes and Codes

LOCAL TIME CHART

Alaska	New York	Gander	London	Cairo	Bombay	Bangkok	Tokyo
-9	-5	-3 1/2	UTC	+2	+5 1/2	+7	+9

TEMPERATURE CONVERSION

°C →	°F °C	← °F	°C →	°F °C	← °F
-40.0	-40	-40.0	-7.8	18	64.4
-37.2	-35	-31.0	-7.2	19	66.2
-34.4	-30	-22.0	-6.7	20	68.0
-31.7	-25	-13.0	-6.1	21	69.8
-28.9	-20	-4.0	-5.6	22	71.6
-26.1	-15	5.0	-5.0	23	73.4
-23.3	-10	14.0	-4.4	24	75.2
-20.6	-5	23.0	-3.9	25	77.0
-18.3	-1	30.2	-3.3	26	78.8
-17.8	0	32.0	-2.8	27	80.6
-17.2	1	33.8	-2.2	28	82.4
-16.7	2	35.6	-1.7	29	84.2
-16.1	3	37.4	-1.1	30	86.0
-15.6	4	39.2	-0.6	31	87.8
-15.0	5	41.0	0.0	32	89.6
-14.4	6	42.8	0.6	33	91.4
-13.9	7	44.6	1.1	34	93.2
-13.3	8	46.4	1.7	35	95.0
-12.8	9	48.2	2.2	36	96.8
-12.2	10	50.0	2.8	37	98.6
-11.7	11	51.8	3.3	38	100.4
-11.1	12	53.6	3.9	39	102.2
-10.6	13	55.4	4.4	40	104.0
-10.0	14	57.2	5.0	41	105.8
-9.4	15	59.0	5.6	42	107.6
-8.9	16	60.8	6.1	43	109.4
-8.3	17	62.6	6.7	44	111.2

To convert from:	To Coordinated Universal Time
Eastern Standard Time	Add 5 Hours
Eastern Daylight Time	Add 4 Hours
Central Standard Time	Add 6 Hours
Central Daylight Time	Add 5 Hours
Mountain Standard Time	Add 7 Hours
Mountain Daylight Time	Add 6 Hours
Pacific Standard Time	Add 8 Hours
Pacific Daylight Time	Add 7 Hours

LAYOVERS